TABLE OF CONTENTS

DISCLAIMER AND TERMS OF USE AGREEMENT:

Introduction – To Humans Time is Linear; To Animals Time Is Life

Chapter 1 – The Interactive Nature of Time

Chapter 2 – Time Is Quantum Physics

Chapter 3 – Time is Either Your Best Friend or Your Worst Enemy

Chapter 4 – Too Late For Fruit; Too Soon For Flowers

Chapter 5 – I Walk a Crooked Line

Chapter 6 – Summary & Conclusion

I Have a Special Gift for My Readers

Meet the Author

The Power of Time
The Interactive Nature of Time
©Copyright 2013 by Dr. Leland Benton

DISCLAIMER AND TERMS OF USE AGREEMENT:

(Please Read This Before Using This Book)

This information is for educational and informational purposes only. The content is not intended to be a substitute for any professional advice, diagnosis, or treatment.

The author and publisher of this book and the accompanying materials have used their best efforts in preparing this book.

The author and publisher make no representation or warranties with respect to the accuracy, applicability, fitness, or completeness of the contents of this book. The information contained in this book is strictly for educational purposes. Therefore, if you wish to apply ideas contained in this book, you are taking full responsibility for your actions.

The author and publisher disclaim any warranties (express or implied), merchantability, or fitness for any particular purpose. The author and publisher shall in no event be held liable to any party for any direct, indirect, punitive, special, incidental or other consequential damages arising directly or indirectly from any use of this material, which is provided "as is", and without warranties. As always, the advice of a competent legal, tax, accounting, medical or other professional should be sought where applicable.

The author and publisher do not warrant the performance, effectiveness or applicability of any sites listed or linked to in this book. All links are for information purposes only and are not warranted for content, accuracy or any other implied or explicit purpose. No part of this may be copied, or changed in any format, or used in any way other than what is outlined within this course under any circumstances. Violators will be prosecuted.

This book is © Copyrighted by ePubWealth.com.

Introduction – To Humans Time is Linear; To Animals Time Is Life

Time! From the dawn of man upon this earth 6,000 years ago, the concept of time has fascinated thinking man and continues even into this age.

In secular circles, time is viewed through quantum physics as having physical properties as well as non-physical properties.

In non-secular circles, time is viewed as "chronos" (Greek for physical time) and "kairos," which are qualitative rather than quantitative. It is time as *a moment*, a significant occasion, and an immeasurable quality. In the New Testament, kairos is God's time, it is real time—it is the eternal now.

In this book we are going to study the concept of time and how it applies to YOUR life. This will be an adventure into an area you never knew existed and it will be fun.

So to start, read the following article. I think you will begin to see the adventure...

Thought Provoking Time

The dominating time-piece is nothing if not thought-provoking. British inventor John Taylor's "Chronophage" (literally 'time eater' from the Greek chronos and phageo) keeps watch outside Cambridge's Taylor Library of Corpus Christi College.(1) A foreboding metal grasshopper with an ominous chomping mouth appears to devour each minute with eerie pleasure and constancy. The toll of the hour is marked by the clanging of a chain into a tiny wooden coffin, which then slams shut—the sound of mortality, says Taylor.(2) The pendulum also speeds up sporadically, then slows to a near halt, only to race ahead again as if somehow calculating the notion that time sometimes flies, sometimes stands still. The invention, according to Taylor, is meant to challenge our tendency to view time itself as we might view a clock. "Clocks are boring. They just tell the time, and people treat them as boring objects," he added. "This clock actually interacts with you"—indeed, striking viewers with the idea that time is nothing to take for granted. (3)

The Christian worldview is one that recognizes at the deepest level that something about humanity is not temporal. Easter, in fact, is the celebration that this is not just a suspicion, but a reality. Christians believe in eternal dwellings, a day when tears will be no more, and in one who is preparing a house of rooms and welcome. (4) And yet, we also very much live with the distinct experience of these promises within time. Christ is not merely the one who will be with us in all eternity, the one who will dry our eyes at time's end. We believe he is also alive and among us today, welcoming a kingdom that is both present and approaching. "Remember, I am with you always," ends one of account of the life of Jesus, "even to the end of the age" (Matthew 28:20). For the Christian, all of time is filled with the hope of resurrection, even as it is filled with Christ himself.

Why then, I wonder, are there moments when time seems so oppressive, the hope of eternity a distant glimmer, and the presence of a resurrected Christ beside the daily pendulum an inapplicable promise? If the Christian life is about moving closer and closer to the glory of the resurrected Christ, why is there not more light and less darkness, a more vibrant Church and less grumbling, greater outreach and less greed, followers who look more like Jesus and less like the world around them? The expectation in the life of a Christian is that there will be a dramatic difference, or at least steady progression, of lives transformed by Christ. But instead we often find little difference—or we find the opposite of progression. Only last week I turned to a friend and asked with a sigh of weariness, "Where is transformation as all this time marches onward?"

John Taylor's menacing grasshopper is an apt image for such a confession. Time marches on oppressively, unapologetically, while the promise of "being transformed into [Christ's likeness] from one degree of glory to another" seems to remain a distant mirage. (5) Christians begin to doubt. Skeptics point out the obvious fantasy of faith. But perhaps something in Taylor's clock also challenges this fearful view of time and transformation. Time is indeed a linear progression, marching onward in precise increments, but our experience of time is far less like this. We are at times startled by its passing, other times painfully aware of its tedious movement. We interact with time knowing that some minutes are fuller than others, but that time is always more than a linear, monotonous experience.

Similarly, when I think of transformation, I often think of dramatic change: an acorn turned into an oak tree, the apostle Paul changed from zealous tormentor to zealous Christian, Lazarus moved from death to life. And I believe there is indeed something quite like this that takes place in the life of one willing to follow resurrected Christ—a creature who actually stops being one thing in order to become something else. It should not be surprising that around the world we find Christians in the most unlikely places, administering aid, speaking hope, exhibiting this change of which the gospel speaks. For clothed in Christ's perfect nature, the nature of a person is truly changed. Though we stand before God imperfect and discouraged, it is the Son the Father now sees. And this part of Christian transformation is as dramatic as it is complete, allowing us—and the world—to stand assured of God's work within.

But this is not to say that God is finished working. To the one who has been united with Christ, the daily indwelling of God is a gift! Within the Christian's experience of time, the message of the gospel is all the more transformational, Christ is our moral influence daily, and through the Holy Spirit we are being further transformed into his image. This kind of transformation is neither the dramatic change often expected, nor the steady linear progression for which we might hope. Like Paul himself, we can find ourselves doing the things we don't want to do, falling back into mindsets that need to be renewed, imitating a broken world more than we imitate Christ. Transformation at these times seems far less like Lazarus rising from the grave and more like a would-be butterfly refusing to come out of its cocoon.

But even here, Christ is surely near, the eternal urging the world of souls to see the potential in this very moment: "The intermediate hope—" writes N.T. Wright, "the things that happen in the present time to implement Easter and anticipate the final day—are always surprising because, left to ourselves, we lapse into a kind of collusion with entropy, acquiescing in the general belief that things may be getting worse but that there's nothing much we can do about them. And we are wrong. Our task in the present... is to live as resurrection people in between Easter and the final day."(6)

That is to say, Easter is being implemented. Whether we make our bed in the depths, whether we fall repeatedly or seem to be moving backward, God is both near and at work, the reality of the resurrection working its way into every ticking minute. In the experience of time before us is the radical promise of both the intermediate hope and transformation and the gift of looking glory full in the face. By the power of the Spirit, God takes the most wretched of creatures and changes it into the likeness of Christ, the most beautiful creature. Whether time is flying or standing still, for the worst of us, even menacing grasshopper types, this is indeed very good news.

Jill Carattini is managing editor of A Slice of Infinity at Ravi Zacharias International Ministries in Atlanta, Georgia.

(1) Maev Kennedy, "Beware the time-eater: Cambridge University's Monstrous New Clock," The Gaurdian, September 18, 2008.
(2) Robert Barr, "Fantastical New Clock Even Tells Time," MSNBC, September 19, 2008.
(3) Ibid.
(4) Luke 16:9, Revelation 21:4, John 14:2.
(5) 2 Corinthians 3:18.
(6) N.T. Wright, Surprised by Hope (New York: Harper, 2008), 29-30.

*****

Okay, time never stops. The minute you look at your wristwatch, time has moved on and left you behind. Without knowing it or perceiving it, you are left in the moment and time has become fleeting to you.

You never own time! Time owns YOU! By working within the confines of the time parameter, we attempt to live our lives always keenly aware that time is running out and that as we age, we are wearing out too.

The time parameter I am speaking about is called the "Space-Time Continuum" in science and as science attempts to define ourselves within this S\pace-Time Continuum, it raises more questions than it answers.

So, in our quest to discover the significance and benefits of time in our lives, let's us first turn to the subject of "The Interactive Nature of Time".

Chapter 1 – The Interactive Nature of Time

We feeble human beings must operate within the Space-Time Continuum so to begin let's define exactly what this is.

What is a space time continuum?

In 1906, soon after Albert Einstein announced his special theory of relativity, his former college teacher in mathematics, Hermann Minkowski, developed a new scheme for thinking about space and time that emphasized its geometric qualities. In his famous quotation delivered at a public lecture on relativity, he announced that,

"The views of space and time which I wish to lay before you have sprung from the soil of experimental physics, and therein lies their strength. They are radical. henceforth, space by itself, and time by itself, are doomed

to fade away into mere shadows, and only a kind of union of the two will preserve an independent reality."

This new reality was that space and time, as physical constructs, have to be combined into a new mathematical/physical entity called 'space-time', because the equations of relativity show that both the space and time coordinates of any event must get mixed together by the mathematics, in order to accurately describe what we see. Because space consists of 3 dimensions, and time is 1-dimensional, space-time must, therefore, be a 4-dimensional object. It is believed to be a 'continuum' because so far as we know, there are no missing points in space or instants in time, and both can be subdivided without any apparent limit in size or duration. So, physicists now routinely consider our world to be embedded in this 4-dimensional Space-Time continuum, and all events, places, moments in history, actions and so on are described in terms of their location in Space-Time.

Space-time does not evolve, it simply exists. When we examine a particular object from the stand point of its space-time representation, every particle is located along its world-line. This is a spaghetti-like line that stretches from the past to the future showing the spatial location of the particle at every instant in time. This world-line exists as a complete object which may be sliced here and there so that you can see where the particle is located in space at a particular instant. Once you determine the complete world line of a particle from the forces acting upon it, you have 'solved' for its complete history. This world-line does not change with time, but simply exists as a timeless object. Similarly, in general relativity, when you solve equations for the shape of space-time, this shape does not change in time, but exists as a complete timeless object. You can slice it here and there to examine what the geometry of space looks like at a particular instant. Examining consecutive slices in time will let you see whether, for example, the universe is expanding or not.

Spacetime

From Wikipedia, the free encyclopedia
http://en.wikipedia.org/wiki/Spacetime

In physics, **spacetime** (also **space–time**, **space time** or **space–time continuum**) is any mathematical model that combines space and time into a single continuum. Spacetime is usually interpreted with space as existing in three dimensions and time playing the role of a fourth dimension that is of a different sort from the spatial dimensions. From a Euclidean space perspective, the universe has three dimensions of space and one of time. By combining space and time into a single manifold, physicists have significantly simplified a large number of physical theories, as well as described in a more uniform way the workings of the universe at both the supergalactic and subatomic levels.

In non-relativistic classical mechanics, the use of Euclidean space instead of spacetime is appropriate, as time is treated as universal and constant, being independent of the state of motion of an observer. In relativistic contexts, time cannot be separated from the three dimensions of space, because the observed rate at which time passes for an object depends on the object's velocity relative to the observer and also on the strength of gravitational fields, which can slow the passage of time.

In cosmology, the concept of spacetime combines space and time to a single abstract universe. Mathematically it is a manifold consisting of "events" which are described by some type of coordinate system. Typically three spatial dimensions (length, width, height), and one temporal dimension (time) are required. Dimensions are independent components of a coordinate grid needed to locate a point in a certain defined "space". For example, on the globe the latitude and longitude are two independent coordinates which together uniquely determine a location. In spacetime, a coordinate grid that spans the 3+1 dimensions locates events (rather than just points in space), i.e. time is added as another dimension to the coordinate grid. This way the coordinates specify *where* and *when* events occur. However, the unified nature of spacetime and the freedom of coordinate choice it allows imply that to express the temporal coordinate in one coordinate system requires both temporal and spatial coordinates in another coordinate system. Unlike in normal spatial coordinates, there are still restrictions for how measurements can be made spatially and temporally (see Spacetime intervals). These restrictions correspond roughly to a particular mathematical model which differs from Euclidean space in its manifest symmetry.

Until the beginning of the 20th century, time was believed to be independent of motion, progressing at a fixed rate in all reference frames; however, later experiments revealed that time slows at higher speeds of the reference frame relative to another reference frame. Such slowing, called time dilation is explained in special relativity theory. Many experiments have confirmed time dilation, such as the relativistic decay of muons from cosmic ray showers and the slowing of atomic clocks aboard a Space Shuttle relative to synchronized Earth-bound inertial clocks. The duration of time can therefore vary according to events and reference frames.

When dimensions are understood as mere components of the grid system, rather than physical attributes of space, it is easier to understand the alternate dimensional views as being simply the result of coordinate transformations.

The term *spacetime* has taken on a generalized meaning beyond treating spacetime events with the normal 3+1 dimensions. It is really the combination of space and time. Other proposed spacetime theories include additional dimensions—normally spatial but there exist some speculative theories that include additional temporal dimensions and even some that include dimensions that are neither temporal nor spatial (e.g. superspace). How many dimensions are needed to describe the universe is still an open question. Speculative theories such as string theory predict 10 or 26 dimensions (with M-theory predicting 11 dimensions: 10 spatial and 1 temporal), but the existence of more than four dimensions would only appear to make a difference at the subatomic level.

*****

The above articles do a good job of empirically defining the Space-Time Continuum, but if you are like me, I want a definition that I can apply to my life.

I know what is in my past and I can do nothing to change it but I have my future to look forward to and this is something I can determine or can I?

Is Our Future Determined or Free?

Questions of freedom and destiny have been raised in every generation. Do we exercise choice, or has everything already been decided? Is our future determined or free? The resultant mental gymnastics leave many feeling confused and others feeling disappointed. Christians insist that nothing takes God by surprise. On the other hand, Christians throughout the ages reject the kind of fatalism that is seen in some parts of the world.

The problem with the question as it is presented is that it is not nearly difficult enough. In order to truly appreciate the magnitude of what we are discussing, we must first deal with an even greater question. And it is this: Imagine if I were able to stop time right now. What would you be thinking? What would you be feeling? The answer is *nothing*.

In the absence of time, we cannot think or feel or do. Everything is frozen. People sometimes complain that I speak too quickly—the problem being that there is not sufficient time for them to think about what has been said. I always try to cheer myself by saying that at least something has been said for them to think about! But it is a fair criticism because in the absence of sufficient time we cannot think things through. In the absence of time altogether, however, we cannot even begin to think, as there is literally no time to think *in*.

The Christian understanding is that we live and have our existence in a space-time continuum. "[We] belong to eternity stranded in time," observes Michael Card. (1) This also means that before God created there was no time; in other words, time is not co-eternal with God. But Christians also attest that God was a thinking, feeling, doing Being even before God created. Can you imagine a Being who is able to think in the absence of time? Of course not, but the God Christians profess not only exists outside of time, God can think and act in the *absence* of time.

Just reading this may be enough to make us feel overwhelmed. And so it should. Whenever we think about the person of God, we should rightly feel that we have come across something truly awesome. And maybe this is part of the problem. We are not faced with a logical contradiction here. Rather, we are faced with the reality of what it means for God to exist, for God to be *God*. You and I are only able to think in time, and thus, God confronts us with choices: "Choose this day whom you will serve," "choose life" and so on (Joshua 24:15; Deuteronomy 30:19). But God, as Christians believe, who is outside of time, sees all of history stretched out before Him. The problem comes, therefore, when the attempt is made to confine God within time. But this needn't be the case. For Christians, a proper understanding of the tension drives us back both to God's divine nature and to our knees, acknowledging how wonderful God is.

This understanding also helps Christians with the issue of eternal life. Many people when confronted with the idea of eternity find the idea frightening, tedious, or absurd. What could one possibly *do* with all of that time? Once again, the dilemma arises because we are captive both to the passage of time and too small a view of who God actually is. People also then ask: if God truly knows all things, then why did God create knowing that we would experience pain in a fallen world? But here Christians attest that God did not create the world and then think of a plan to rescue it. The book of Revelation depicts that the Lamb was slain before the foundations of the world were laid. This does not mean that the crucifixion took place in our space-time history before creation (there was no space-history for it to take place in). What it does mean is that even before God created, God also knew the cost—the suffering of his own Son—to redeem creation and unite us with the Father. God didn't count that cost too great—and hence Christians sing of God's amazing grace.

Let me conclude with the following. The Christian God is big enough to be able to say, "I know the plans I have for you… plans to prosper you and not to harm you, plans to give you hope and a future" (Jeremiah 29:11). There is no hope without a secure future, and the future is frightening in the absence of hope. Only God is big enough to bring these two things together—hope and a future—and this is what God has done for us.

Michael Ramsden is European director of Ravi Zacharias International Ministries in the United Kingdom.

(1) Lyrics from "Joy in the Journey" by Michael Card, 1986.

*****

The above non-secular article is quite interesting so let's discuss it.

The non-secular point of view is more applicable to life than the scientific or secular view. Science uses a systematic approach to the theories proposed and never takes into consideration the "human factor" of feelings, emotions, wants and needs.

The article makes a good point that without time, we don't exist. Only God can exist outside the Space-Time Continuum.

Now the following is provocative and deep so read it slowly to take it all in please...

Science tells us that there is only one element in the universe that is both mass and energy. This element is light! Everything else is either mass or energy.

In non-secular circles, God is defined as the Light of this world. Which makes sense because God is mass in the bodily form of His son Jesus Christ, and is energy in the form of the Holy Spirit.

Now, man is both mass and energy, which also makes sense since non-secular thought tells us that we were created in the image of God, which means we share is eternal nature.

Now mass, cannot exist outside the Space-Time Continuum, which again makes sense since when we die we leave our earthly bodies behind but our spirits are eternal and can and do leave the Space-Time Continuum.

This is what the article above is attempting to explain. Outside of the Space-Time Continuum, is multi-dimensional. We live in a three dimensional world. It is said that Christ was 11-dimensional. The theories of a multi-dimensional universe abound in science. But one fact is clear; we only exist outside the Space-Time Continuum as energy or spirit.

Einstein stated in his Law of Relativity that mass and energy are neither created nor destroyed. In other words, the universe has a finite supply of mass and energy PERIOD.

Mass is static and remains so since it can neither be created nor destroyed. But energy is slowly being dissipated away as spirits leave the universe.

But this statement has sparked tremendous controversy mostly between secular and non-secular thought with the secular side being the most vocal.

I am not going to discuss the various controversies but leave it to my readers to sort out in their own minds. It really boils down to whether you believe in God or not although both sides couch their arguments to remove God from the equation.

Now let's discuss Time as Quantum Physics...

Chapter 2 – Time Is Quantum Physics

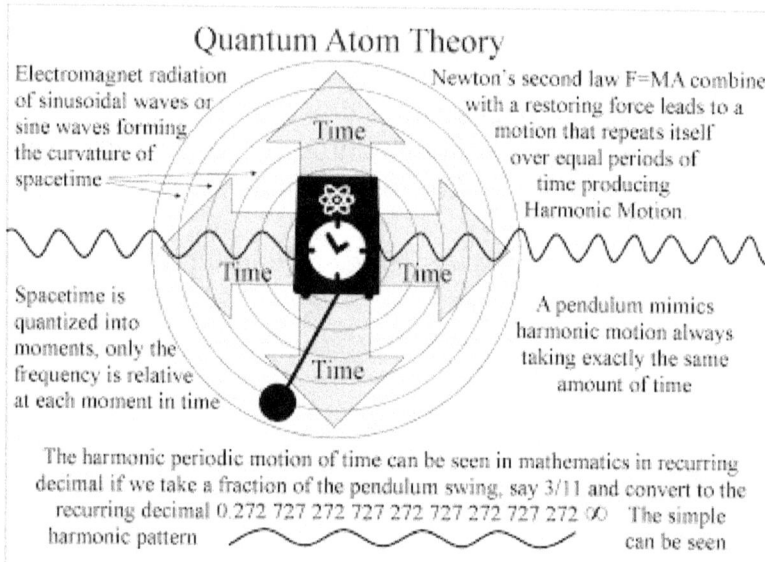

Quantum Physics or "quantum" as it is most often referred to is most often identified with Einstein and his Law of Relativity.

In non-secular thought, the first chapter of Genesis is the foundational chapter of the Bible and, therefore, of all true science. It is the great creation chapter, outlining the events of that first week of time when "the heavens and the earth were finished, and. . . . God ended his work which he had made" (Genesis 2:1-2). Despite what the evolutionists say, God is not creating or making anything in the world today (except for special miracles as recorded in Scripture) because all His work was finished in that primeval week. He is now engaged in the work of conserving, or saving, what He first created.

There are only three acts of special creation--that is, creation out of nothing except God's omnipotent word--recorded in this chapter. His other works were those of "making" or "forming" the created entities into complex, functioning systems.

His first creative act was to call into existence the space/mass/time cosmos (Space-Time Continuum). "In the beginning God created the heaven and the earth" (Genesis 1:1).

This is the domain which we now study in the physical sciences.

The second is the domain of the life sciences. "God created . . . every living creature that moveth" (Genesis 1:21).

It is significant that the "life" principle required a second act of direct creation. It will thus never be possible to describe living systems solely in terms of physics and chemistry.

The third act of creation was that of the image of God in man and woman.

The study of human beings is the realm of the human sciences. Our bodies can be analyzed chemically and our living processes biologically, but human behavior can only really be understood in terms of our relation to God, whose image we share.

What is most significant and what I want you to take from this treatise here is that we are more than a bunch of chemicals and biological processes; we have a mind and spirit that can be co-eternal with God. This is an individual's choice. God created us; we create the quality of our lives by the choices we make.

The next article is a real eye-opener and puts into perspective the differences of time…

Breaking Time

"Uncanny" was one of the vocabulary words on my sixth grade vocabulary list, which was to be found within the book we were reading as a class. I remember thinking Madeleine L'Engle's *A Wrinkle in Time* was exactly that—*uncanny*, peculiar, and uncomfortably strange. Yet I also remember that it stayed with me—the story of a quirky girl named Meg, her overly-intelligent little brother, and their time-transcending journey to save their physicist father with the help of three mysterious beings. L'Engle's book, which recently celebrated its fiftieth anniversary, invites readers to see time itself differently. Her stories will no doubt continue to perplex sixth graders, and stay with us long after we have set them aside.

L'Engle is the writer who first showed me the incredible difference between two words in Greek, which we unfortunately translate identically. To the English reader, *chronos* and *kairos* both appear to us as "time." But in Greek, these words are vastly different. Chronos is the time on your wrist watch, time on the move, passing from present to future and so becoming past. Kairos, on the other hand, is qualitative rather than quantitative. It is time as *a moment*, a significant occasion, and an immeasurable quality. In the New Testament, kairos is God's time, it is real time—it is the eternal now.

Thus, when Jesus stepped into time to proclaim the kingdom of God among us, he came to show us in chronos the reality of kairos. "Jesus took John and James and Peter up the mountain in ordinary, daily chronos," writes L'Engle. "Yet during the glory of the Transfiguration they were dwelling in kairos."(1) With this story in mind, L'Engle describes kairos as that time which breaks through chronos with a shock of joy, time where we are completely unselfconscious and yet paradoxically far more real than we can ever be when we are continually checking our watches. "Are we willing and able to be surprised?" L'Engle asks. "If we are to be aware of life while we are living it, we must have the courage to relinquish our hard-earned control of ourselves."(2) For the Christian, the thought is particularly consequential. We must have the courage to see counterculturally beyond ourselves and our self-importance. We must have the courage to live aware that the kingdom of God is close at hand.

I imagine Jacob discovered this difference between chronos and kairos when he set aside both the past that was about to catch up with him and his paralyzing fear of the future only to find himself living in "none other than the house of God" (Genesis 28:17). The prophets and others describe similar moments of waking to the present and finding the eternal dimensions of time. The shepherds in Bethlehem were going about their ordinary work when an angel appeared before them and the glory of the Lord shone around them. "Do not be afraid," the angel announced. "I bring you good news of great joy that will be for all the people. Today in the town of David a Savior has been born to you" (Luke 2:13-14). At this invasion of kairos into the routine of chronos, the shepherds chose to respond with action: "Let's go to Bethlehem and see this thing that has happened, which the Lord has told us about" (2:15).

In the life, death, resurrection, and ascension of Christ, where eternity steps into time and invites us to see far beyond our watches; we are presented with a similar decision. Are we willing to be surprised again by Christ's coming? Are we willing to act on it? Are we able to release the nervous control of our daily schedules in order to stop and see the resurrected Christ, the eternal now in our midst? With every prophet who proclaimed the coming of the Messiah in history, the apostle calls out today, "Behold, now is the time (kairos) of God's favor, now is the day of salvation" (2 Corinthians 6:2).

Whether we are ready for our sense of time and self to stop at such an invasion, Christ has come. He comes quietly and unexpectedly; he comes and upsets our very notions of time and all we discover within it. The eternal Word steps into flesh, into our bounded realm of time, and literally embodies the reality that time is meaningful because of the eternal one in our midst. His presence reminds us that kairos is breaking into chronos and transforming it, transforming us. It proclaims, "The kingdom of God is close at hand"—and invites us to join the world breaking in along with it. In moments of life and death, present fears and plans for the future, might this radical invasion move us and our vantage points, as we find ourselves continually surprised by the one who comes so near.

Jill Carattini is managing editor of A Slice of Infinity at Ravi Zacharias International Ministries in Atlanta, Georgia.

(1) Madeleine L'Engle, *Walking on Water: Reflections on Faith and Art* (New York: Bantam, 1982), 93.
(2) *Ibid.*, 99.

*****

In presenting both the secular side as well as the non-secular side of the way the concept of time is viewed, I want my readers to see the full argument where science and religion literally "collide" because each presents their argument in such a way so as to be very convincing. A reader must practice discernment and study the arguments carefully in order to decide what makes the most sense.

I am a behavioral scientist for over 32-years so I am trained in the sciences. Many of you reading this are not. Science is such that it can be the great "deceiver" as it couches terms and definitions in language meant to confuse.

I am also trained in religion and hold a doctorate degree from the seminary I attended. It is important to note that religion is man-made and no God-made and many religions are flawed in their presentation and doctrine. In fact, in many ways, they are just as flawed as science and this is why a reader must practice discernment.

When I was growing up, my mentors and my teachers taught me not to follow any man – men change and get it wrong. They taught me to follow God – He never changes and His truth is absolute.

I have the best of both worlds where you – the reader – do not. I am trained in both secular and non-secular thought and ideals. So read my words carefully and be discerning to the max. Don't allow anybody to do your thinking. Do your own thinking on all matters.

Chapter 3 – Time is Either Your Best Friend or Your Worst Enemy

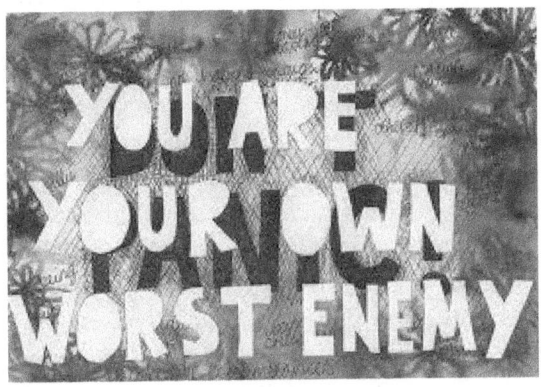

I think everybody on the planet is familiar with the adage, "Time is either your best friend or your worst enemy." But what if your best friend is your worst enemy?

Let me explain...

Here is a list of generations so you can keep them straight in your mind as I explain:

2000/2001-Present - New Silent Generation or Generation Z
1980-2000 - Millennials or Generation Y
1965-1979 - Generation X
1946-1964 - Baby Boomers
1925-1945 - Silent Generation
1900-1924 - G.I. Generation

Baby Boomers - A baby boomer is a person who was born during the demographic post-World War II baby boom between the years 1946 and 1964. The term "baby boomer" is sometimes used in a cultural context. Therefore, it is impossible to achieve broad consensus of a precise definition, even within a given territory. Different groups, organizations, individuals, and scholars may have widely varying opinions on what constitutes a baby boomer, both technically and culturally. The red segment from 1946 to 1964 is the postwar baby boom. Baby boomers are associated with a rejection or redefinition of traditional values; however, many commentators have disputed the extent of that rejection, noting the widespread continuity of values with older and younger generations. In Europe and North America boomers are widely associated with privilege, as many grew up in a time of widespread government subsidies in post-war housing and education, and increasing affluence. As a group, they were the wealthiest, most active, and most physically fit generation up to that time, and amongst the first to grow up genuinely expecting the world to improve with time. They were also the generation that received peak levels of income; therefore they could reap the benefits of abundant levels of food, apparel, retirement programs, and sometimes even "midlife crisis" products. One feature of Boomers was that they tended to think of themselves as a special generation, very different from those that had come before. In the 1960s, as the relatively large numbers of young people became teenagers and young adults, they, and those around them, created a very specific rhetoric around their cohort, and the change they were bringing about. This rhetoric had an important impact in the self perceptions of the boomers, as well as their tendency to define the world in terms of generations, which was a relatively new phenomenon.

Generation X - commonly abbreviated to Gen X, is the generation born after the Western post-World War II baby boom. Demographers, historians and commentators use beginning birth dates from 1964-1980. Generation X encompasses the 44 to 50 million Americans born between 1965 and 1980. This generation marks the period of birth decline after the baby boom and is significantly smaller than previous and succeeding generations.

Generation Y - If you were born from 1980-2000 you are a member of Generation Y. With numbers estimated as high as 70 million, Generation Y (also known as the Millennials) is the fastest growing segment of today's workforce. Millennials grew up with easy access to computers, constant connection to the internet, and smart phones in-hand. They're accustomed to using technology in every part of their lives and fully believe in its power to make their lives easier. You could even say that they're intolerant of companies and individuals that aren't equally as savvy. This generation dreams without limits. They're in search of two things – money and happiness. Wonder why you receive resumes with five employers in less than five years? This is why! Members of Generation Y are constantly searching for the next best thing and when they find it (or think they find it) they'll jump on it! Instead of passing notes and calling a friends' house, millennials grew up trading text messages and chatting online. They're used to communicating quickly without ever saying a word. This is why they'll rapidly respond to text and email but stare at the phone for days giving themselves a pep talk before returning a call. A life of instant gratification has created an extremely impatient generation of people. They're quick to spend and intolerant of wait times.

I live in a college town in Southern Utah so I enjoy chatting with the local college kids at Starbucks. These kids represent Generation Z or what has been labeled 'The Lost Generation" (see the chart above).

Each generation faces its own unique set of problems but Generation Z got hit with a double whammy – 9/11/01 and the recession and economic fallout of 2008.

They are living in a time of deep inopportunity. They are coming out of the universities with no job prospects and no hope.

Many adult children are still living with their parents. This fact coupled with the fact of being burdened with high student loan debt is causing many kids to give up. They can't get a job; they can't afford to have their own apartment and living quarters; they can't pay off their student loans, etc. They seem to be stuck in a "time warp".

Time is their worst enemy! Every year the universities and colleges continue to graduate new students adding more kids to the workforce problem.

These kids see no hope; time has left them behind. Their parents are helpless to assist them in overcoming their problems. Oh, they help out financially and provide living quarters but this is only a band-aid.

The world is changing too rapidly; the family unit as we know it is under assault from many directions. Time has seemingly become the enemy and is no friend…or has it?

Problems are like pain. They force us to focus and pay attention when we would rather ignore them and forget them. Unfortunately they don't go away on their own.

Inaction is death; doing nothing and ignoring the problems will cause a person to become a mentally depressed individual. And many of these are just that – depressed into inaction as they play their video games and withdraw from relationships and society.

Abraham Lincoln once said, "You have the right to complain if you have the heart to help."

This is how I attacked the problem…

I went to the local university and got permission to teach a job fair called "How to Find a Job in Today's Marketplace: Are your skill sets marketable?"

The response was overwhelming with over 2500 kids signing up for the symposium on the first day.

On the first day we broke the kids up into accountability teams based on their skill sets and college degrees. Each kid filled out an extensive questionnaire and each accountability team had sic kids – 3-women and 3-men.

We then identified job skills that were in demand. The local university here has one of the finest nursing schools in the country and nurses are in demand. However; there are four other fine nursing schools within 100-miles of the local university's nursing school so there is a glut of nurses in Southern Utah. The problem is that the graduating nurses do not want to move away from their family and friends. They are in love with the familiar.

I brought in hospital HR people from around the country to sit with these nursing kids and make them offers to move. A good many of these offers contained continuing education packages which are important in nursing. A graduating nurse with a bachelor's degree in nursing is in demand but nurses with Masters Degrees walk on water with nail scars in their hands. The HR people were very persuasive in their offers and the nurses were given the option of having their moving expenses paid to accept the position and if they weren't happy the hospital would pay their moving expenses back to Utah. Needless to say, a good many accepted the offers and moved to new jobs.

So, the first step was to teach the kids to identify their skill sets to see if they were marketable. There isn't a good deal you can do with a history degree or art degree so these kids need to learn skill sets that were marketable first. We accomplished this by identifying job openings that were closely compatible with their degrees and interests as defined by the questionnaire. I brought in experts from these industries to teach them what skills were needed for them to gain employment.

Next, I taught them the value of networking using social media and following experts that could provide them with job opportunities and developing marketable skill sets. They were very good at social media but needed direction on how to take advantage of it.

I emphasized industries that were hiring and one such industry was my own company ForensicsNation. I have over 22,000 forensic investigators around the world. We hire interns and teach them how to find bad guys on the Internet and candidly the kids jumped at this opportunity because forensic investigators are in demand and are never without work. Plus the fact computers are a way of life for this generation and they could work at home.

I brought in other forensic companies to hire interns and hire they did. Never had these companies had such a huge source of college degreed kids to choose from for their individual intern programs.

From Southern Utah, I branched out hiring Job Fair teams to go to universities around the country copying our success in Southern Utah.

Here is my point: inaction leads to apathy and depression. We have enough of that already. The goal is "action!" Even if it wrong do something and then correct it. Get up you dead backside and make time work for you!

MAKE TIME WORK FOR YOU!

I recall a conversation I had with an inmate recently released form 10-years of prison. I expected to interview an angry man but I was surprised. In the course of the conversation, I learned that this man was banned from the securities industry for his crime of securities fraud but spent the 10-years in prison retraining himself to be a court reporter. When I asked him why he did this, he looked at me and said, "I had a choice to make; I am young and when I am released I knew I would have to find a job to support myself. **I didn't serve time; time served me!"**

Amazing!

Chapter 4 – Too Late For Fruit; Too Soon For Flowers

In the last chapter the inmate made a statement, "I didn't serve time; time served me!"

This chapter is called "Too Late For Fruit; Too Soon For Flowers," and what this means is that time has stuck you in the middle or in limbo. It is too late for the beginning and too soon for the ending so what is a person to do within this new time parameter?

How does a person get time to serve them? Well let's talk about this now…

Time is in constant motion and no one can keep up with the passage of time. But what we can do is act on what is called the "compartments of time".

Using the inmate cited above, he could easily have served out his prison time playing cards, working out on the weight pile, etc. like the other inmates do but he realized that he needed a new career when he was released and he acted upon this within the compartment of time given him.

We cannot keep up with time but we can ACT within time compartments. In Chapter 3, I described college kids that we languishing because of no jobs. I showed them how to act within their individual compartments of time and by so doing their lives improved.

This time compartments I am speaking about come at us daily. In my book, "The Power of trained Observation" http://www.amazon.com/dp/B00BSRYMGW I teach people how to identify and evaluate these time compartments and whether to accept them of discard them.

Opportunities abound and come at you daily but if you do not recognize them as opportunities they are lost forever.

For many people, they are truly "stuck" in limbo and they don't even know it. I find this to be very sad, especially when action breaks the lethargy of inaction.

In today's economic downturn, it is so easy to give up and go on government unemployment. I have witnessed many groups of people using the excuse that "they deserve it' to become professional unemployed people with many refusing to look for jobs or refusing to take any job that is not equivalent to the one they lost. This is not "real world" people; the world has changed and left them by the roadside.

As we have witnessed recently in the news. The "free ride" is coming to an end as unemployment insurance runs out leaving a good many people scrambling to find income. They served time and time didn't serve them.

I am familiar with a good many cases where the individuals rather than looking for a job instead have been searching for any and all aid from the state, local city governments and church groups in order to keep their free ride going rather than work. These people have become professional slackers!

Serving time never offers any benefits; having time serve you is where the benefits lie and where your life changes. You must act within the compartments of time granted to you. You only have choices and not excuses!!!

The interactive nature of time cannot be denied. Action promotes more action and solves problems whatever they are. You have the power to choose action and you have the free will to choose so let's discuss free will since this one subject really confuses people.

Free Will

Much experimentation has occurred on the subject of free will. Does free will exist and does it act on the brain via attention? I cite Ben Libet's now famous experiment in which subjects were asked to flex their wrists whenever they chose while being monitored by devices on their scalps, which were designed to measure the readiness potential, associated with preparation for movement. On average, the readiness potential was detected 550 milliseconds before movement occurred; however, not all readiness potentials were followed by movement. When subjects were asked to report when they were conscious of deciding to move, it was found that awareness preceded the movement by 100 to 200 milliseconds. Some experts claim that this casts free will into doubt, as the signal to move came before awareness of the desire to move.

Libet posits that contrary to this disproving the existence of free will it is evidence that we are able to block the preparation process and stop movement from occurring altogether, since awareness still comes before movement. Put this way it would seem as if much of the will we exert is aimed to stop subconscious urges, a concept called "free won't."

Libet believes that free will is not just the power to veto unbidden urges, though. He claims that free will is an active force.

Pointing to a plethora of experimental data, he says that willful mental effort is correlated with activity in the prefrontal cortex.

Repetitive (remember most learning is repetitive), automatic acts do not induce high amounts of activity in this area; whereas tasks involving concentration do (concentration or focus is a very important concept we will discuss more later on).

Schizophrenics show unusually low activity in this area, corresponding to lack of control. Patients with lesions to the prefrontal cortex respond reflexively to environmental cues, performing actions with no thought. From the experiments cited, it seems clear that this area of the brain is necessary for willful action, which Libet says in turn influences the brain.

<u>What is all-important to this theory is attention</u>. Selective focusing of attention filters out distractions and makes one concentrate on one or a few particular elements present in sensory input. Although the data one receives is the same whether one is paying attention or not, the brain's response to the data is changed. We can decide to some extent what information registers and what does not. Thus, our perceptions are changed. Or, to put it another way, mental force affects the activity of the brain in a perceptible way. Given that we can change brain activity by paying attention to something, thus affecting the rate at which corresponding sets of synapses fire, it follows that attention is important for neuroplasticity (Neuroplasticity is the lifelong ability of the brain to reorganize neural pathways based on new experiences. As we learn, we acquire new knowledge and skills through instruction or experience. In order to learn or memorize a fact or skill, there must be persistent functional changes in the brain that represent the new knowledge. The ability of the brain to change with learning is what is known as *neuroplasticity.*).

After all, as a certain neural pathway fires more and more often it becomes stronger and even more likely to be activated again. Just as sensory input can change cerebral cortex organization, PET scans have shown that attention can too (remember this fact because it is extremely important). Libet even goes so far as to say that our mental states shape what we perceive more than the original stimulus does.

WOW! Did you understand this last comment? Our perceptions shape our responses and actions more than the reality or original stimuli!!! Reality versus perception again rears its ugly head! Our mental states – the life force of focusing – shape what we perceive more than the original stimulus.

In other words, we are what we perceive and reality takes a back seat. **But, we observe completely when our focus is away from ourselves.**

This is very similar to "Zen" philosophy but more pure. Zen is about observation of "what is." Now, the 'what is' is the 'what is' *below* your filters and assumptions. In other words, the goal and point of Zen is to watch your mind as it meaning-makes, while not attaching to either the process or your assumptions. In this way, the mind sort of slides to the side, and you peer past the mind-games to the essential 'emptiness' of everything. (I don't believe this is true because Zen implies detachment and emptiness and I cannot abide either one of these. But what is certain, focusing your attention away from you will cause you to observe everything around you.)

Let's take our discussion a little bit farther for just a moment by discussing what is true or not true and how we arrive at truth. I need you to see that your eyes and ears really do see and hear everything around you BUT, the mind deploys filters based on a variety of reasons that effectively block your awareness of the various stimuli.

Faith in Observation and the Nature of Truth by Zach Gildersleeve

The role of the observer in science is both supremely important and overwhelmingly critical. The observer has really come into its own through the science of the twentieth century; dominating physics - the traditional and classical "hard" science - once objective, now subjective just as is anthropology or psychology. This historical progression is neatly summed up in Michael Frayn's Copenhagen:

We put man back at the center of the universe (mankind is constantly trying to become the center of the universe). Throughout history we keep finding ourselves displaced (fallen mankind)...to the periphery of things. First...a mere adjunct of God's unknowable purposes...then we're pushed aside again by the products of our own reasoning! Until we come to the beginning of the twentieth century and we're suddenly forced to rise from our knees again... [Einstein] shows that measurement - measurement, on which the whole possibility of science depends - measurement is not an impersonal event that occurs with impartial universality...it's a human act.

My response - Any human act is subjective by nature and never objective or absolute. Science must always be absolute and completely objective.

The character of physicist Niels Bohr then goes on to describe how Einstein's thinking led his colleagues to develop the modern theories of indeterminism and uncertainty, in which "...the universe only exists as a series of approximations. Only within the limits determined by our relationship with it. Only through the understanding lodged inside the human head."

My response - Unfortunately this is seriously flawed. Although we dream of the perfect we do live in the flawed! The universe is not a subset of ourselves; we live in the universe. If we view anything only within the framework or limits determined by our relationship with it, we are left with an incomplete picture and/or a corrupted viewpoint.

I think this is brilliant prose, and I completely agree with the optimistic tone that the passage carries. I fully believe that the observer is what makes science possible.

My response - Another flawed assumption when you believe that man is at the center of the universe and in control of his own destiny. This is like saying that your wife exists only to fall in love with you and if you didn't love her she wouldn't exist.

Without the observer, there can be no observations.

My response - This is true; however with or without the observer it doesn't mean there is nothing to observe.

I reject a platonic worldview where we are only observing reflections of Truth, reflections of a reality that would exist whether or not we exist.

My response - Oh really; a typhoon rages in the Philippines however I am in Las Vegas. Do you really want me to believe that it is not raging because I am not there?

On the contrary, I think that without the observer, without the participant, nothing would exist. This argument is echoed in the Strong Anthropic Principle, although I think that the universe exists because we see it, rather than the universe existing and consequently so then must we.

My response - Good try but no prize...this to me is just double-speak.

Observations make up the base of science, and they have been there since science was created from the logical processes of Aristotle. During the scholasticism of the Middle Ages, observations were decentralized, but even until they were restored to their central importance by the Scientific Revolution of the 16th and 17th centuries, they never disappeared. They make up the data and the empiricism of modern science. Without observations, and therefore without the observer, science would not be possible.

Of course, this is a relatively straightforward understanding, something that would be found at the beginning of any popularized science book, or in the introduction of most science textbooks. Something Carl Sagan would have discussed in Cosmos. I am interested in the why: Why can we observe things? Why do things exist to observe in the first place? Why, when we see phenomena, do we infer that there must be a before and an after? Why are we able to make connections between different observations? These are difficult questions, and I think very relevant to the Origins class because of their importance in science. If observations were impossible to make, or if it were impossible to share them among others, science would also be impossible. You, as a person exist, even if I never meet you or observe you. If you did not know about this course and had never participated, this course would continue.

The universe is rational...and therefore observers are possible (although keep in mind that I feel that the visa versa of this statement is more true.) I have yet to find answers for these questions, and I think that even if we can or will be able to answer the how we may never know the why. The ability to make observations is fully a priori knowledge (*a priori* knowledge is independent of experience) in that it must be possible for other knowledge to occur. I can never prove that what I am observing exists outside of the constructs of my mind, because the mind filters all the senses and always creates just as much as it observes. This is very true!!!

An interesting case study of this is called blindsight, in which patients with damage to their primary visual cortex loose the consciousness of sight. The patient will insist they are blind, and yet will be able to find their way around unfamiliar settings, and respond to purely visual stimuli. Blindness hones the other four senses and makes them more acute.

Other similar conditions exist where patients suffering from hysterical deafness will respond to the cry of a child, or those born with an extraordinarily magnetic sense of direction that, when blindfolded and lost, can easily find their way home. In all these cases, the mind filters the observational input. In addition to being a challenge to behaviorism in claiming that there is a very important connection between consciousness and the seemingly "subconscious" behavioral responses, cases like blindsight act as a nail in the coffin of the objective-worldview.

The mind is a very subjective piece of matter through which all observations are filtered. The nihilist and pragmatist in me believes that only things that can be fully known, understood, or reasoned can exist, but this belief is confounded by the apparent complete lack of an objective truth in the face of an active mind. It is impossible to speak of the why concerning the mind and the nature of observation because such a question would at best still be filtered through the very mind we hope to comprehend.

Given this, I think it is evident that our ability to observe (the why, at least) boils down to a question of faith, a faith that all of science is founded on. Faith is a word that is normally reserved for religion, used when talking about things that cannot be proven...not a word usually reserved for science. However, like the postulates of geometry and the a priori knowledge of epistemology, the faith in observation at the heart of science is not going away, nor could it hope to be separated from the scientific method. However, the scientific method does supplement the fallible human senses, exactly as Francis Bacon intended. His methodology gives us a structure with which we can compare and share our observations, and trust that the data other people observe is what we would see as well (although still keeping the conclusions we draw from the observations a more personal thing and more open to disagreement.)

In summation, the observations every individual makes are completely subjective, a personal truth, but when a large enough group is sampled, the truths fall into a curve, where it is reasonable to assume the highest point is the Truth. The bulk of the remainder of my paper will attempt to understand if this Truth can only be considered a scientific truth stemming from scientific observations, or if this same probabilistic method can produce truth from all areas of thought; art and religion in particular. In essence, what is truth? As I have previously disclosed, the only truth is objective or absolute truth since it never changes. Scientific truth is not objective truth insofar as man interprets it differently. True, there are never-changing scientific truths. I am not speaking about these here.

Observation is the faith of science, the theory upon which science is made possible. It has been postulated by some social scientists that essentially the reverse of this statement is true; that faith is an observation.

They speak of the faith that exists in religious experience, but I think the observation of faith can include a large amount of human experiences, such as the appreciation of beauty or of art, or from any synthesis of emotion such as love, jealously, or hatred that seems to be emergent in the human psyche and above "base" responses to stimuli. I agree...mostly. Just as observation is the faith of science, faith is the observation of religion. However, they go on to say that those faith experiences could be considered a 6th sense, something beyond but equal to the Aristotelian five. I disagree. Faith and emotional responses are products of the mind, not of the external world

In cinema, there has never been any doubt that the audience responds differently accordingly depending on the emotional synthesis created by the audience, not the film itself. A film alone is a length of celluloid or tape. It has no inherent emotional content, only what the spectator adds. This was first understood systematically by the Russians, primarily Lev Kuleshov, a pioneering father of film editing as we know it today. The famous Kuleshov effect, an interesting psychological experiment, demonstrates how much emotion the audience projects on a piece of art. Kuleshov created a short film in which a close up of a famous actor was juxtaposed (preceding in this case) with three different images - a bowl of soup, a dead woman in a coffin, and a girl playing with a toy bear. Audience members reported the actor as feeling hunger, sorrow and joy respectively, even though the image of the actor was exactly the same across all three. The different shots acquire their meaning only from their relation to other shots and how we come to understand these relationships...exactly what the Bohr character in Copenhagen spoke of.

Therefore, it is important to keep in mind (no pun intended) that the constructs of the mind are exactly that...constructs. As anyone who has been moved by a film, a piece of art, or a religious service can attest to, it is still a very powerful experience, but it is not based purely on data the way that the 5 senses are under the scientific method. As each individual has a different mind and a different set of filters, a methodology is needed to create some sort of common language that everyone can speak and share. I would argue that with any religious or emotional response, this methodology is missing and consequentially there is no "language" with which we can communicate these experiences unless they are translated in a common spoken or written language. In the process of translation, something is always lost.

Of course, we can talk about when we cried, or what made us fear, but in translating these emotions into the spoken language the intention changes. In fact, only when such an experience is translated back to the subjective is the emotion regained. Thus, if a painting makes me feel cold and blue, a more accurate way to describe my emotions to another human would be to create another piece of art intending to make others cold and blue rather than simply to talk about my experience using everyday language. The problem is, in religion, that there is no way to retransmit the emotions that people attribute to a god or creator (love, awe, etc.) because we are unable to reconstruct a truly divine presence in ourselves.

To pass on art-inspired emotions through art makes us artists, but to pass on religious-inspired emotions would make us gods.

But isn't this exactly what Christianity teaches? I would argue that the production of all forms of art, from the very beginning, stems from either a literal or figurative quest for immortality; the desire to create something permanent or to affect others, to record accomplishments and life. This same passion for immortality inspires religious devotion, but to communicate it through a further religious experience would be impossible because it would be a man-made construct, and not something wholly-other. Religious faith is therefore a completely subjective and internal experience

It also is difficult to speak about religious faith using the model of probabilistic truth mentioned earlier. It might be possible to create a normal curve of religious beliefs and take the highest point as the Truth, but keep in mind that many of these beliefs are associated with second hand experiences, the transmission that I spoke of earlier, and such experiences cannot be regarded as God, only as our understanding of God. Thus, using probability it would be possible to create a Truth regarding how humans perceive God, but this would speak nothing about the Truth of God.

Observation as a faith and faith as an observation are two mutually exclusive methods of understanding the world around us. If they are put together, we are thrown into an infinite loop, as there is no solid foundation upon which to build logic. Indeed, this is true of all thought...the best and most necessary room for faith is at the base, the leaps of faith that make further thought possible. After that point, thought is just a matter of logical progression. I have no reason to think that God would need any faith beyond faith in God alone, faith in the existence of God. God would have to be supremely rational and would not need faith beyond a priori theology.

Miracles would not be leaps of faith, but logical progressions. I fully believe that for a God to exist and exact worship or respect from God's creations, God would create a fully rational world. Such a world would not need miracles or any more than a single leap of faith to experience God. In my mind, miracles are proofs against the existence of God because of their irrationality and because they require constant and new leaps of faith. I completely reject any model of God, as does Kenneth Miller, of God-the-Charlatan, any God that would create irrational behavior, something that when considered as a probabilistic Truth would be irrational. To do so would be a trick, a conscious [sic] illusion on the part of God.

In my view the model of probabilistic Truth offered earlier can only be applied to science, because only with science can the subjective and often distorted observation and mind processes be supplemented with a methodology that speaks the same direct language as the observations. Science is the best method we have to understand observations, to understand our relationships to them and through the understanding lodged in the human mind. It is the only way that we can come close the existence of a universal Truth. However, there are many subjects that science will never be able to broach, and will be unable to create any Truths. Are we then to accept that Truth can only exist where science can find it; otherwise is there no Truth?

What then is the definition and nature of Truth and truths? Truth (with a capital T) is one of those elusive things that are difficult to nail down precisely, like complexity or life. However, I think that the preceding arguments can point to a more clear understanding of what I view as Truth. There are two options. The first being if there is a Truth, then it is entirely personal, because even objective observations of the natural world are processed in the mind and are therefore altered until they become subjective. Truth is personal, and there can be no hope for a universal Truth, excepting when two personal Truths are coincidentally the same. The other option is that Truth can be communicated via a translation process like the scientific method in science or the creation of more art in the artistic process. In this case, there is no Truth, only individual truths that can be freely shared and expressed, but are limited by their relativism and again, the subjective experience of the observer.

It seems to me that both options speak of the irrationality of an objective Truth, because such a Truth would be "perverted" by the constructs of the mind, and wholly incommunicable to boot. Personally I like the idea of a personal Truth, rather than subjective truths, because of the ramifications: I can create Truth from the sheer power of my mental processes; it makes me god. As a god who knows the Truth I can control my destiny and be confident in my decisions and future. And I think I could get use to that. .

*****

Okay, everything so far has been meant to lay a solid foundation. We are what we perceive and these perceptions are filtered through underlying belief systems accumulated over our lifetimes.

Many of these belief systems can be quite dangerous while others are good and protective.

Things, habits, and beliefs are LEARNED and can be unlearned just as easily as they were learned in the first place; however some things learned are not easy to give up. We call these addictions. The human mind gravitates toward depravity and all things sensuous and feeling. And the mind does not give up anything sensuous! Trained observation is recognizing all stimuli in the conscious mind first and evaluating it in the conscious mind prior to subconscious perception.

This evaluation is based on focus, which we will discuss in detail very soon. Example: a soldier in a combat situation in a jungle sees a bug but discards it immediately because a bug cannot hurt him. He sees a tree and focuses on this because an enemy sniper can be in the tree. A leaf is discarded…a vine is discarded but a rock and pond is focused upon because both can be used for concealment. Evaluation using focus prequalifies what we see prior to subconscious thought. The mind strives to think so it is important that we prequalify what it is we think about.   Now let's put "Choice" into perspective

Chapter 5 – I Walk a Crooked Line

The following is taken from the Introduction of my book, "I Walk a Crooked Line" http://www.amazon.com/dp/B00CF9FXW4.

> *"Consider the work of God: for who can make that straight, which he hath made crooked? All things therefore whatsoever ye would that men should do unto you, even so do ye also unto them: for this is the law and the prophets. Enter ye in by the narrow gate: for wide is the gate, and broad is the way, that leadeth to destruction, and many are they that enter in thereby. For narrow is the gate, and straitened the way, that leadeth unto life, and few are they that find it."*

This is not like any book you have ever read you have read before. It is about you and it is about me. The internal battle we fight daily is between good and evil within us. We struggle to live a peaceful quiet life but the battle is never far away.

From the least of us to the greatest, the battle is present daily and as we live our lives, we struggle to do good and not evil. One of the greatest men who ever lived – the Apostle Paul - put it this way…

"For we know that the law is spiritual: but I am carnal, sold under sin. For that which I do I know not: for not what I would, that do I practice; but what I hate, that I do. But if what I would not, that I do, I consent unto the law that it is good. So now it is no more I that do it, but sin which dwelleth in me. For I know that in me, that is, in my flesh, dwelleth no good thing: for to will is present with me, but to do that which is good is not. For the good which I would I do not: but the evil which I would not, that I practice. But if what I would not, that I do, it is no more I that do it, but sin which dwelleth in me. I find then the law, that, to me who would do good, evil is present. For I delight in the law of God after the inward man: but I see a different law in my members, warring against the law of my mind, and bringing me into captivity under the law of sin which is in my members. Wretched man that I am! Who shall deliver me out of the body of this death?"

*****

The Apostle Paul was without a doubt one of the greatest Christians that ever lived, and in the 1st century Palestine, evil was ever present and eventually took his life in Rome.

Here is an interesting article I think will enlighten you…

Prayer for a Future Me

While thoughts and resolutions for the year ahead are crossing many of our minds, Matt Sly and Jay Patrikios are still thinking 30 years into the future. Sly and Patrikios are the minds behind the 2002 website "Future Me" that allows people to send messages to themselves years or decades from the time they were written. In the year 2015, a man named Adam is set to get an e-mail from himself that asks, "Do you still write? Do you still draw? Does Radio Shack still exist?" Sly explains the rationale: "We want people to think about their future and what their goals and dreams and hopes and fears are. We're trying to facilitate some serious existential pondering."(1)

A quick overview of some of the publicly-posted messages shows people doing just that. Some are pondering dreams they hope to have accomplished by the time they hear from themselves in the future: "I hope you are moving up in your job... I also hope you are making more responsible choices." Others are taking it as a moment to remind themselves what they were up to years earlier or record what they hope will be beyond them in the future: "I hope you're better because as I'm writing this letter, you're doing terrible." It is a time capsule wrought in an e-mail, readily drawing in participants all over the world. At the very least, it extracts in many a sense of intrigue. At most, sending words to future selves seems to draw a sense of nostalgia, accountability, apprehension, or hope.

I used to keep a journal that mostly held thoughts and events consumed with present days. I seemed most prone to write in it when something was happening or had just happened, when something was on my mind or on my heart *at present*. But there is one page far in the back that differentiated from the others. In scattered sentences now crammed on a page full of thoughts I speak to days far ahead of me: "Remember that you wanted to be the kind of woman that grows old gracefully." "If you ever become a parent, I hope you will be the kind who can say 'I'm sorry.'" "When it's time to let go of certain freedoms, take it with poise." "If it's ever your turn to face disease, remember that you wanted to do it with faith; you wanted death never to scare you more than resurrection gives you hope." While I like to think of these mental notes as prayers for the future—and many of them are—many of them more closely resemble a listing of fears, an anxious warning at what I *might* forget or what *might* go wrong. Though I am looking ahead, it is as if I am still looking behind me.

In an essay titled "Please Shut This Gate" English author F.W. Boreham describes signs carefully placed by landowners throughout the landscape of New Zealand. "Please shut this gate," was a message one could read often throughout his countryside, signs placed by fence owners intent on keeping some things from wandering away and some things from wandering in. Depicting this common scene, Boreham then draws a parallel to the importance of shutting similar gates in our own lives, closing the door that keeps things both in and out. He writes, "[W]hen Israel escaped from Babylon, and dreaded a similar attack from behind, the voice divine again reassured them. 'I, the Lord thy God, will be thy rearguard' (Isaiah 58:8). There are thousands of things behind me of which I have good reason to be afraid; but it is the glory of the Christian evangel that all the gates may be closed. It is grand to be able to walk in green pastures and beside still waters unafraid of anything that I have left in the perilous fields behind me."(2)

Whether looking down roads to the New Year or the coming decades, it is the gift of the follower of Jesus that there are gates that may be closed. We need not worry about the future, nor look to resolutions or future me's with fear of failing, nor tremble at what Christ has put behind us—or in front of us. In the words of a seventeenth century Puritan: "To suppose that whatever God requireth of us we have power of ourselves to do is to make the Cross and grace of Jesus Christ of none effect."(3) Christ has written a message across the future to be delivered to our laboring souls each new day. As he went head first into the shadows of self-giving, he cried, "It is finished," forever offering a door to shut, forever promising the strength to shut it. In this New Year, one can say in hope and in light: Christ has gone before us, he walks among us, he is our rearguard, and he is our strength.

Jill Carattini is managing editor of A Slice of Infinity at Ravi Zacharias International Ministries in Atlanta, Georgia.

(1) Matt Sly and Jay Patrikios, Futureme.org.
(2) F.W. Boreham, "Please Shut This Gate," *The Silver Shadow* (New York: The Abingdon Press, 1919), 118-119.

(3) John Owen, *Works of John Owen: Volume 3* (Edinburgh: T & T Clark, 1862), 433.

*****

I have read some of the messages posted on "Future Me" and they all possess dreams of hope of what the person can be. But will they be? Ah, this is one of the questions I will attempt to answer in this book.

Now read this article…

The Best Intentions

How far can we get on good intentions? According to one survey conducted among a diverse group of men and women, thirty percent of those who make New Year's resolutions admit not keeping them into February. Just one in five continues his or her resolution for six months or more. Apparently, we don't get very far.

<u>We meet life with intentions to succeed, intentions to be a good person, intentions to live life to the fullest.</u> Yet however many ways we might interpret success, goodness, or full-living, our good intentions have certain aspects in common: the hope to improve, the idea of becoming something more than what we are at the moment, the expectation that one should reach his or her potential. It is as if there is an image implanted in our minds that upholds the idea of something we *could* be or *might* be—some even use the language of even being *meant* to be. But there is all too often a tragic side to best intentions. When they are not fully realized, there is usually a sense that it is we who have gotten in the way.

Great minds from Augustine to G.K. Chesterton saw clearly that the most verifiable truth of the Christian worldview is certainly the depravity of humanity. It can be observed across countries and languages, at any time and within every decade, from barbaric accounts of depravity in faraway places to more accepted forms of depravity close at home. We close our eyes to reality where we refuse to see the same story repeating itself again and again. We might euphemize the thought of sin into neurotic myth, outdated opinion, or church propaganda, but it has not been euthanized. Observe for a short time at any playground and you will note quickly amongst even the youngest that something is amiss. If we were to truly observe our hearts, motives, and wills, we would hardly find them good and consistent leaders to follow.

The Christian worldview recognizes the recurring story of a disappointed and disappointing humanity. Not only do we miss our own intentions, we miss the intention of one we faintly recognize within us; we sense in our createdness the greater mark and glory of the creator disappointingly out of reach. The one who spoke to the dejected Eve in the Garden of Eden and to the defiant David through the prophet Nathan is the present one beside whom we, too, stand in contrast. We can step no closer to that standard by our own intentions than a foolish king can order the stars to bow before him. To look at the Son is to find that even our *best* intentions are made of straw.

Yet looking at Christ, we not only see our humanity beside a perfect human, we find this perfect human moving toward us in mercy, giving us a bigger picture of the good and the fullest, and ushering us into the possibility of holding more than we ever imagined. Where we are honest about our limits and shortfalls, we can truly grasp the beauty of Jesus and the unimaginable depth of a Father's love. It is in Christ where we find that God moves the blur of sin to give us the picture of all God intended. And here, we find the Christian worldview not only coherently offers the diagnosis, but also the cure.

The late Christian songwriter Rich Mullins alluded to the bigger pictures of God when he observed of his own life: "What I'd have settled for You've blown so far away/ What You brought me to/ I thought I could not reach." In the intentions of God, we find that where we would have settled, where we would have been content with success or goodness, the Father moves us far beyond. Where we would have fallen beyond reach, the Son took our place. "God who is mighty," proclaims the psalmist "has done great things for me." In the coming of this New Year, might we recognize a similar story in our own lives?

Jill Carattini is managing editor of A Slice of Infinity at Ravi Zacharias International Ministries in Atlanta, Georgia.

*****

The best intentions seem to scurry away from us. We try but fail often to be a good person; evil always seems to lurk nearby; failure is evil's best friend. But does it have to be so? Read on please...

Coming Home

There is a line in the story of the prodigal son that is easy to miss. It comes as the transition in the story, but it also seems to mark the transition in the son. The story is familiar. Not long after the younger son demands the right to live as he pleases, after he leaves with his father's money and gets as far away as possible, and after he loses everything and is forced to hire himself out in the fields, the story reads that the prodigal "came to himself" and, at this, he decides to turn back to the father.

Today it is often translated that the son "came to his senses," as we might describe a man or woman who, on the precipice of a bad decision or impulsive act, decides to turn around. But the phrase in the Greek literally describes the prodigal as *coming to himself*, and seems to point at something far more than good decision-making. In a sermon titled "Bread Enough and to Spare," popular English preacher Charles Spurgeon notes that this Greek expression can be applied to one who comes out of a deep swoon, someone who has lost consciousness and comes back to himself again. The expression can also be applied to one who is recovering from insanity, someone who has been lost somewhere within her own mind and body, only to come back to herself once again.

With both of these metaphors, the son is one who wakes to health and life again, having been unconscious of his true condition. Standing in a foreign field hungry and alone, the son comes to something more than a good decision. He is waking to an identity he knew in part but never fully realized. He is remembering life in his father's house again, though for the first time.

Human identity seems a succession of inquiry and wakefulness. For some of us, who we are is discovered in layers of life and realization, questioning and consciousness. Essayist Annie Dillard articulates this progression of awareness and the rousing of self as something strangely recognizable—"like people brought back from cardiac arrest or drowning." There is a familiarity in the midst of our awakenings. We wake to mystery, she writes, but so somehow we wake to something *known*.

The Christian tells a similar story of waking to life in the most fully human sense of the word. We are like those who have lost consciousness, caught in the madness of our own condition, longing to be released, until we are awakened to life despite ourselves with one so eager for our homecoming. The apostle concurs:

"You were dead through the trespasses and sins in which you once lived, following the course of this world, following the ruler of the power of the air, the spirit that is now at work among those who are disobedient... But God, who is rich in mercy, out of the great love with which he loved us even when we were dead through our trespasses, made us alive together with Christ."(1)

Coming to ourselves, we wake to human need, to human condition, to our poverty and our dignity, claiming in our very identities our need for resurrection, our need for home.

One further use of this expression comes out of the old world fables of enchantment. With this metaphor, "coming to ourselves" is like coming out of a magician's spell and assuming once again our true forms. It is reminiscent of the scene in *The Silver Chair* where the children are trapped beneath Narnia in the land called Underworld and persuaded to believe there is no such thing as a Narnian. The Queen of Underworld, who is really a witch, has thrown a green powder into the fire that produces a sweet and drowsy smell. In this enchanting haze, their identity as Narnians becomes hazy, and the world they thought they knew begins to disappear. But it is at this moment of despair that Puddleglum makes a very brave move. With his bare foot he stomps on the fire, sobering the sweet and heavy air. "One word, Ma'am," he says coming back from the fire, limping, because of the pain. "Suppose we *have* only dreamed or made-up, all those things... Suppose this black pit of a kingdom of yours *is* the only world. Well, it strikes me as a pretty poor one... We're just babies making up a game, if you're right. But four babies playing a game can make a play-world which licks your real world hollow... I'm on Aslan's side, even if there isn't any Aslan to lead it. I'm going to live as much like a Narnian as I can even if there isn't any Narnia. So, thanking you kindly for our supper, we're leaving your court at once and setting out in the dark to spend our lives looking for Overland."

Coming out of their enchantment, the prisoners of Underland remembered they were children of another kingdom. Coming to themselves, they began to realize who they were all along. What if waking to our identities as children of the Father is like uncovering the people God has created us to be from the start? What if coming to ourselves is like remembering we are citizens of a better kingdom, a kingdom we vaguely recall and yet long to return? The prodigal's awakening came as the startling recognition that there was plenty in his father's house, and that he himself was starving. Waking to this, we reclaim the very identities given to us in the beginning. And doing so, we come to ourselves because we are setting out for home again. We come to ourselves because we are going to the Father.

Jill Carattini is managing editor of A Slice of Infinity at Ravi Zacharias International Ministries in Atlanta, Georgia.

(1) Ephesians 2:1-5.

*****

Coming to ourselves is something we need to ponder for it is this very act of seeking that takes us to the answer.

Who among us can live a life without seeking something? It is patently impossible to live a life without direction and seeking the right direction. Most people seek the "good life' but what is the good life? Read this and ponder...

Ancient Wisdom for Good Living

"What is the good life?" is a question as old as philosophy itself. In fact, it is the question that birthed philosophy as we know it. (1) Posed by ancient Greek thinkers and incorporated into the thought of Socrates through Plato, and then Aristotle, this question gets at the heart of human meaning and purpose. Why are we here, and since we are here, what are we to be doing? What is our meaning and purpose?

Out of the early Greek quest for the answer emerged two schools of thought. From Plato emerged rationalism: the good life consists of ascertaining unchanging ideals—justice, truth, goodness, beauty—those "forms" found in the ideal world. From Aristotle emerged empiricism: the good life consists of ascertaining knowledge through experience—what we can perceive of this world through our senses. (2)

For both Aristotle and Plato, rational thought used in contemplation of ideas is the substance of the good life. Despite the obvious emphasis by both on goodness emerging from the contemplative life of the mind (even though they disagreed on the source of rationality) both philosophers saw the good life as impacting and benefiting society. For Plato, society must emulate justice, truth, goodness, and beauty, so he constructs an ideal society. For Aristotle, virtue lived out in society is the substance of the good life, and well-being arises from well-doing.

Not long ago, I conducted an internet search on the tag "What is the good life?" and I was amazed at what came up as the top results of my search. Most of the entries involved shopping or consumption of one variety or another. Some entries were on locations to live, and still others involved self-help books or other media aimed at helping one construct a good life. Others were the names of stores selling goods to promote "the good life." There were no immediate entries on Plato, Aristotle, or the philosophical quest that they helped inaugurate. There were no results on wisdom or the quest for knowledge lived out in a virtuous life. Instead, most of the entries involved material pursuits and gains. Sadly, this reflects our modern definition of what is good.

Perhaps, what are for many individuals still very trying economic times; it is difficult not to equate material items with the good life, more money, more security, or more opportunity. While it has always been said of every generation that these are times of great crisis and upheaval, we feel this search for meaning anew and afresh today, and perhaps wonder at the practicality or wisdom of looking to the past for insight or understanding into the good life.

And yet, the ancients remind us that "not even when one has an abundance does one's life consist of possessions" (Luke 12:15). Abundant or meager as they may be, possessions must not make up the substance of one's life. Instead, their proper use necessarily involves right living in community. Perhaps the ancient Hebraic wisdom is particularly instructive in a time in which we might equate goodness with what we possess. "He has told you, *what is good*; and what does the Lord require of you but to do justice, to love kindness, and to walk humbly with your God?" (Micah 6:8) This vision of the good life, cast not when times were good, but during a time when calamity and exile awaited the nation of Israel offers an alternative understanding. Do justice, love kindness, and live out both of those virtues in light of humility before God; this is what is good and is the ground of the good life.

The wisdom of the ancients, from the Greeks and the Hebrews, suggests that the good life can be attained regardless of circumstance or possession. It shimmers in the wisdom of justice and kindness. It is found in the application of knowledge rightly applied in relationship to the world around us. It shines in humility before the God who is *good,* and is part and parcel of a relationship with that God. The good life is not bought or sold; it is not a prime real estate location, or a formula for success. The good life is our life offered to God and to others in justice, kindness, and humility.

Margaret Manning is a member of the speaking and writing team at Ravi Zacharias International Ministries in Seattle, Washington.

(1) A.L. Herman, *The Ways of Philosophy: Searching for a Worthwhile Life* (Scholars Press: Atlanta, 1990), 1.
(2) *Ibid*, 82.

*****

So now the lesson is over; I have done my best to teach you what I know but let me take it one step further and outline for you what you have learned....

Chapter 6 – Summary & Conclusion

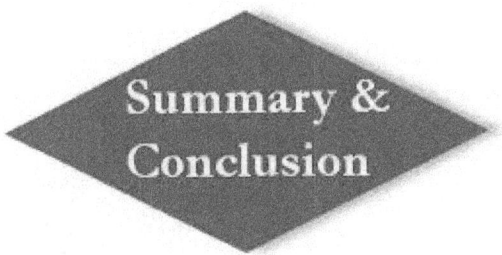

Let's summarize what you have learned…

In the Introduction, I describe time as linear to humans but not to animals. Animals use time to guide their lives – in reproduction, in migration, in navigation and more. Humans use time as tool to track events, the movement of stars, day-night condition, history, and more also.

In secular circles, time is viewed through quantum physics as having physical properties as well as non-physical properties.

In non-secular circles, time is viewed as "chronos" (Greek for physical time) and "kairos," which are qualitative rather than quantitative. It is time as *a moment*, a significant occasion, and an immeasurable quality. In the New Testament, kairos is God's time, it is real time—it is the eternal now.

In Chapter 1, I described the Interactive Nature of Time and defined the Space-Time Continuum. I went on to say that outside the Space-Time Continuum, we do not exist. Matter or mass cannot exist outside of time.

Only energy can exist outside of time and our spirits are energy so our spirits are eternal.

In Chapter 2 I defined time as quantum physics. I defined the non-secular view of creation as being three acts of god. His first creative act was to call into existence the space/mass/time cosmos (Space-Time Continuum). "In the beginning God created the heaven and the earth" (Genesis 1:1). This is the domain which we now study in the physical sciences. The second is the domain of the life sciences. "God created . . . every living creature that moveth" (Genesis 1:21). It is significant that the "life" principle required a second act of direct creation. It will thus never be possible to describe living systems solely in terms of physics and chemistry. The third act of creation was that of the image of God in man and woman.

In Chapter 3, I defined time in the context that "Time is Either Your Best Friend or Your Worst Enemy". I used the example of the job fair that I threw for the local university and made it clear that action was required rather than inaction.

In Chapter 4, my goal was to introduce to you the concept called "Too Late For Fruit; Too Soon For Flowers," and what this means is that time has stuck you in the middle or in limbo. It is too late for the beginning and too soon for the ending so what is a person to do within this new time parameter? Time is in constant motion and no one can keep up with the passage of time. But what we can do is act on what is called the "compartments of time".

In Chapter 5 I dfined choice and fee will and how "I Walk a Crooked Line," which is the tile of onec of my books. We only have choices and not excuses!!!

72

Okay, I hope you enjoyed my words and that they have caused you to sit diwn and have a "good think".

Now I have a spoecial gift for you…read on.

I Have a Special Gift for My Readers

I appreciate my readers for without them I am just another author attempting to make a difference. If my book has made a favorable impression please leave me an honest review. Thank you in advance for you participation.

My readers and I have in common a passion for the written word as well as the desire to learn and grow from books.

My special offer to you is a massive ebook library that I have compiled over the years. It contains hundreds of fiction and non-fiction ebooks in Adobe Acrobat PDF format as well as the Greek classics and old literary classics too.

In fact, this library is so massive to completely download the entire library will require over 5 GBs open on your desktop.

Use the link below and scan all of the ebooks in the library. You can select the ebooks you want individually or download the entire library.

The link below does not expire after a given time period so you are free to return for more books rather than clog your desktop. And feel free to give the link to your friends who enjoy reading too.

I thank you for reading my book and hope if you are pleased that you will leave me an honest review so that I can improve my work and or write books that appeal to your interests.

Okay, here is the link…

http://tinyurl.com/special-readers-promo

PS: If you wish to reach me personally for any reason you may simply write to mailto:support@epubwealth.com.

I answer all of my emails so rest assured I will respond.

Meet the Author

Dr. Leland Benton is Director of Applied Web Info, a holding company for ePubWealth.com, a leading ePublisher company based in Utah. With over 21,000 resellers in over 22-countries, ePubWealth.com is a leader in ePublishing, book promotion, and ebook marketing.

As the creator and author of "The ePubWealth Program," Leland teaches up-and-coming authors the ins-and-outs of today's ePublishing world. He has assisted hundreds of authors make it big in the ePublishing world.

Leland also created a series of external book promotion programs and teaches authors how to promote their books using external marketing sources.

Leland is also the Managing Director of Applied Mind Sciences, the company's mind research unit and Chief Forensics Investigator for the company's ForensicsNation unit. He is active in privacy rights through the company's PrivacyNations unit and is an expert in survival planning and disaster relief through the company's SurvivalNations unit.

Leland resides in Southern Utah.

Visit some of his websites
http://appliedmindsciences.com/
http://appliedwebinfo.com/
http://BoolbuilderPLUS.com
http://embarrassingproblemsfix.com/

http://www.epubwealth.com/
http://forensicsnation.com/
http://neternatives.com/
http://privacynations.com/
http://survivalnations.com/
http://thebentonkitchen.com
http://theolegions.org

www.ingramcontent.com/pod-product-compliance
Lightning Source LLC
Chambersburg PA
CBHW071800170526
45167CB00003B/1107